# Ecosystems
# Estuaries

**Simon Rose**

www.av2books.com

AV² provides enriched content that supplements and complements this book. Weigl's AV² books strive to create inspired learning and engage young minds in a total learning experience.

## Your AV² Media Enhanced books come alive with...

 **Audio** Listen to sections of the book read aloud.

 **Key Words** Study vocabulary, and complete a matching word activity.

 **Video** Watch informative video clips.

 **Quizzes** Test your knowledge.

 **Embedded Weblinks** Gain additional information for research.

 **Slide Show** View images and captions, and prepare a presentation.

 **Try This!** Complete activities and hands-on experiments.

Go to www.av2books.com, and enter this book's unique code.

**BOOK CODE**

**W606113**

AV² by Weigl brings you media enhanced books that support active learning.

**...and much, much more!**

---

Published by AV² by Weigl
350 5th Avenue, 59th Floor
New York, NY 10118
Website: www.av2books.com    www.weigl.com

Copyright ©2014 AV² by Weigl
All rights reserved. No part of this publication may be reproduced, stored in a retrieval system, or transmitted in any form or by any means, electronic, mechanical, photocopying, recording, or otherwise, without the prior written permission of the publisher.

Library of Congress Cataloging-in-Publication Data
Rose, Simon, 1961-
 Estuaries / Simon Rose.
     pages cm. -- (Ecosystems)
 Includes index.
 ISBN 978-1-62127-485-8 (hardcover : alkaline paper) -- ISBN 978-1-62127-488-9 (softcover : alkaline paper)
 1. Estuarine ecology--Juvenile literature. 2. Estuarine biology--Juvenile literature. 3. Estuaries--Juvenile literature. I. Title.
 QH541.5.E8R67 2013
 577.7'86--dc23
                          2012044726

Printed in the United States of America in North Mankato, Minnesota
1 2 3 4 5 6 7 8 9 0   17 16 15 14 13

042013
WEP300113

**Project Coordinator** Aaron Carr
**Design** Mandy Christiansen

Every reasonable effort has been made to trace ownership and to obtain permission to reprint copyright material. The publishers would be pleased to have any errors or omissions brought to their attention so that they may be corrected in subsequent printings.

Photo Credits
Weigl acknowledges Getty Images as its primary photo supplier for this title.

# Contents

AV² Book Code . . . . . . . . . . . . . . . . . . 2
What is an Estuary Ecosystem? . . . . . . . . 4
Where in the World? . . . . . . . . . . . . . . 6
Mapping Estuaries . . . . . . . . . . . . . . . 8
Estuary Climates . . . . . . . . . . . . . . . . 10
Types of Estuaries . . . . . . . . . . . . . . . 12
Estuary Features . . . . . . . . . . . . . . . . 13
Life in Estuaries . . . . . . . . . . . . . . . . 14
Estuary Plants . . . . . . . . . . . . . . . . . 16
Estuary Mammals . . . . . . . . . . . . . . . 18
Estuary Birds, Reptiles,
Fish, and Crustaceans . . . . . . . . . . . . 20
Estuaries in Danger . . . . . . . . . . . . . . 22
Science in Estuaries . . . . . . . . . . . . . . 24
Working in Estuaries . . . . . . . . . . . . . . 26
Underwater Viewer . . . . . . . . . . . . . . 28
Create a Food Web . . . . . . . . . . . . . . 29
Eco Challenge . . . . . . . . . . . . . . . . . 30
Key Words/Index . . . . . . . . . . . . . . . 31
Log on to www.av2books.com . . . . . . . . 32

# What is an Estuary Ecosystem?

Estuaries are often gathering places for migratory birds.

Many different forms of life inhabit the Earth, each with their own unique living requirements. These **organisms** are dependent on their environment in order to survive. As well, all animals and plants in an environment depend upon each other. They are interconnected. This system of interactions between organisms and their environments is called an **ecosystem**.

An estuary is a body of water where fresh water from rivers mixes with the salt water of the ocean. An estuary may also be called a bay, harbor, inlet, **lagoon**, or sound. Estuaries are located anywhere rivers or streams enter the ocean.

Ecosystems | **Estuaries**

The mixture of salt water and fresh water that occurs in estuaries creates a unique ecosystem. Estuaries are home to a variety of organisms, many of which might not normally interact with each other outside of this environment.

### Eco Facts

In the United States, more than 110 million people live in coastal regions. This puts a great strain on estuary ecosystems.

## Levels of Organization in Estuary Ecosystems

Ecosystems can be broken down into levels of organization. These levels range from a single organism to many **species** of organism living together in an area.

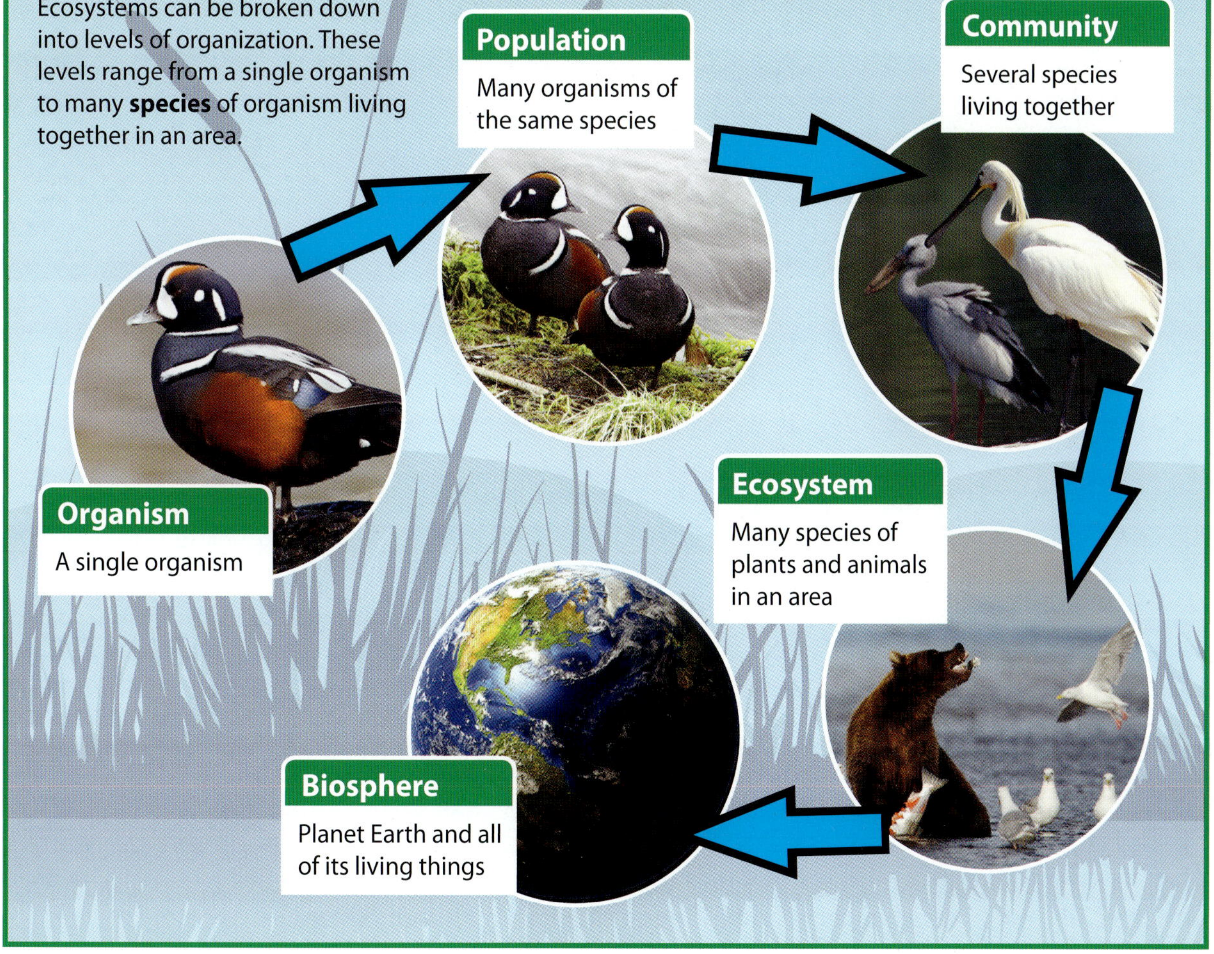

**Organism**
A single organism

**Population**
Many organisms of the same species

**Community**
Several species living together

**Ecosystem**
Many species of plants and animals in an area

**Biosphere**
Planet Earth and all of its living things

# Where in the World?

The Gulf of St. Lawrence is home to large populations of harp seals.

There are estuaries in all parts of the world. Several of these estuaries are very large. The world's largest estuary is the Gulf of St. Lawrence in Eastern Canada. It is through this estuary that the water from the Great Lakes empties into the Atlantic Ocean. This estuary contains about 8,397 cubic miles (35,000 cubic kilometers) of water in an area of 91,000 square miles (235,689 square kilometers).

The Ob River in Western Siberia flows into the Arctic Ocean. This river empties into the Gulf of Ob, which is the world's longest estuary. It is about 500 miles (805 kilometers) long and around 50 miles (80 km) wide.

Ecosystems | **Estuaries**

In South America, the Amazon River has a drainage basin of about 2.7 million square miles (7 million sq. km). All of the water in this basin drains into the Amazon estuary. The Amazon enters the Atlantic Ocean in an estuary that is about 150 miles (241 km) wide. The Amazon estuary has many islands and extends inland for more than 190 miles (306 km).

Also in South America, the estuary of the Río de la Plata forms part of the border between Argentina and Uruguay. This estuary is 180 miles (290 km) long and 140 miles (225 km) wide where it enters the Atlantic Ocean.

Chesapeake Bay on the Atlantic coast is the largest estuary in the United States. More than 150 rivers drain from a basin measuring 64,299 square miles (166,534 sq. km). Chesapeake Bay is 200 miles (322 km) long and 30 miles (48 km) at the widest point.

### Eco Facts

The city of London is on the Thames River, a few miles from the Thames estuary. Pollution from this city threatens the Thames estuary downstream. In recent years, much work has been done to reduce the amount of pollution entering the water system.

| Fishing is an important industry on Chesapeake Bay. |

# Mapping Estuaries

This map shows where some of the world's largest estuaries are located. Find the place where you live on the map. Do you live close to an estuary? If not, which estuary is closest to you?

### Legend
- Estuaries
- Ocean
- River
- Land

**Scale at Equator**

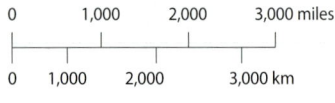

N

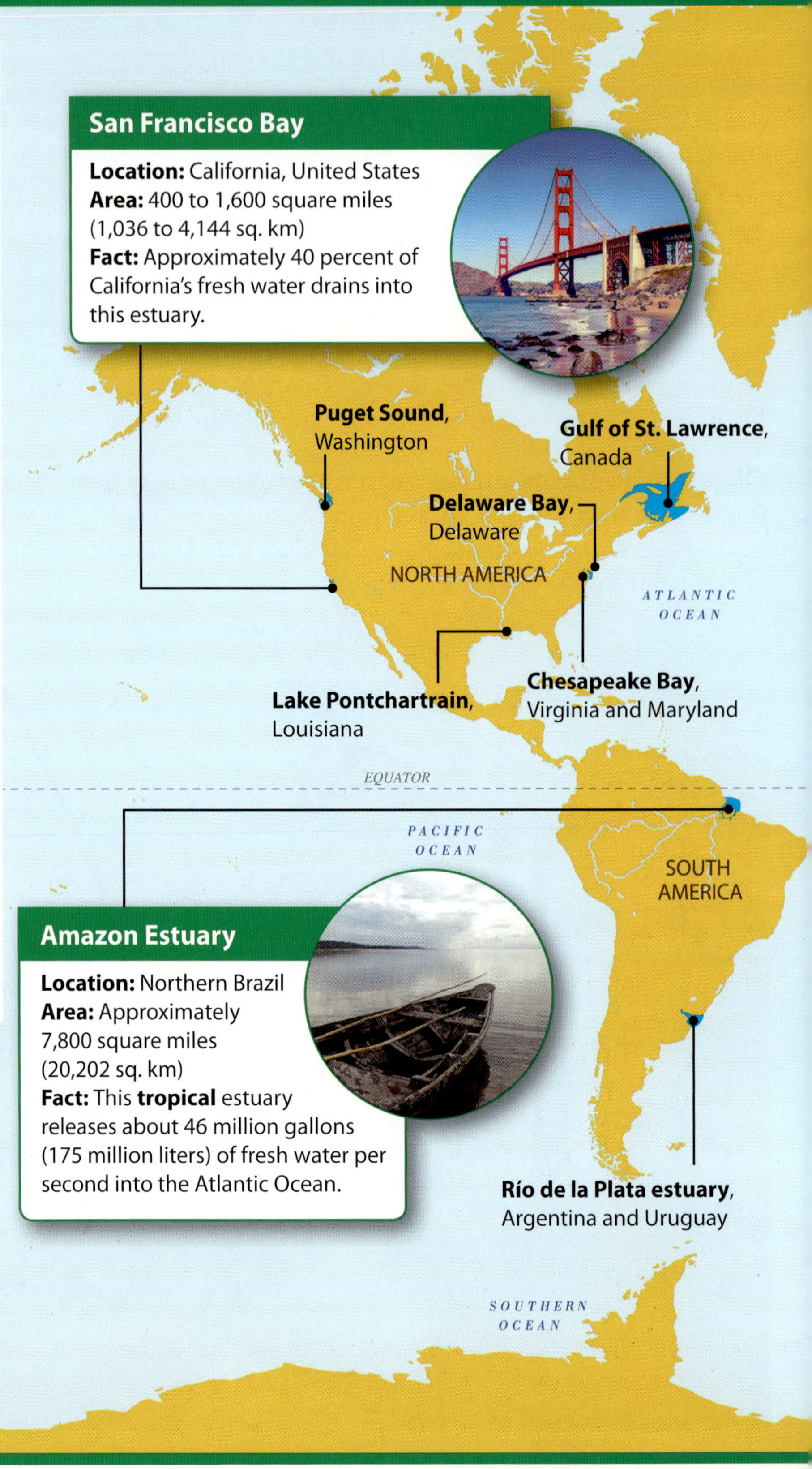

### San Francisco Bay
**Location:** California, United States
**Area:** 400 to 1,600 square miles (1,036 to 4,144 sq. km)
**Fact:** Approximately 40 percent of California's fresh water drains into this estuary.

**Puget Sound**, Washington

**Gulf of St. Lawrence**, Canada

**Delaware Bay**, Delaware

NORTH AMERICA

ATLANTIC OCEAN

**Lake Pontchartrain**, Louisiana

**Chesapeake Bay**, Virginia and Maryland

EQUATOR

PACIFIC OCEAN

SOUTH AMERICA

### Amazon Estuary
**Location:** Northern Brazil
**Area:** Approximately 7,800 square miles (20,202 sq. km)
**Fact:** This **tropical** estuary releases about 46 million gallons (175 million liters) of fresh water per second into the Atlantic Ocean.

**Río de la Plata estuary**, Argentina and Uruguay

SOUTHERN OCEAN

Ecosystems | **Estuaries**

**Firth of Forth**, Scotland

**Fjords**, Norway coast

**Ob River estuary**, Western Siberia, Russia

ASIA

ARCTIC OCEAN

### Thames Estuary

**Location:** United Kingdom
**Area:** 323 square miles (837 sq. km)
**Fact:** The Thames is one of the largest estuaries in Great Britain and a major shipping route.

EUROPE

**Loire estuary**, France

**Tagus estuary**, Portugal

AFRICA

**Gambia estuary**, West Africa

PACIFIC OCEAN

INDIAN OCEAN

AUSTRALIA

**Spencer Gulf**, South Australia

**Kaipara Harbour**, New Zealand

SOUTHERN OCEAN

ANTARCTICA

9

# Estuary Climates

The Gulf of St. Lawrence is ice covered in winter, though it does not freeze solid.

The climates of the world's estuaries vary greatly, from the frozen Arctic to the warm waters near the **equator**. An estuary's climate depends on where it is located. The temperature of the seawater greatly affects what kinds of organisms can thrive in a particular estuary system.

## Estuaries in Colder Regions

The Ob and the Gulf of St. Lawrence have cold climates. The River Ob flows into the Russian Arctic. The average January temperature is −18° Fahrenheit (−28° Celsius) and 40°F (4°C) in July. The estuary's water freezes solid in the winter. The Gulf of St. Lawrence climate is more moderate. The average temperature in winter is 14°F (−10°C), and 59°F (15°C) in summer. In winter, icebergs are not uncommon in the Gulf of St. Lawrence.

Ecosystems | **Estuaries**

## Tropical Estuaries

The Amazon estuary is located at the equator and has only two seasons, the rainy season and the dry. The rainy season occurs between January and June. During this time, some areas may receive more than 87 inches (221 centimeters) of rain. The average temperature in the rainy season is 86°F (30°C). During the dry season, which occurs between July and December, total rainfall ranges from 4 to 8 inches (10 to 20 cm). The average temperature during the dry season is about 90°F (32°C).

### Eco Facts

Due to their varied geographic locations and climates, different plants and animals thrive in each of the world's estuaries. This makes each estuary ecosystem in the world unique.

## Temperate Estuaries

These estuary types have temperate, or milder, climates with four distinct seasons. Chesapeake Bay is a temperate estuary climate. It enjoys hot and humid summers and mild to cold winters. Although many of the rivers that flow into the bay may freeze in winter, it is rare for Chesapeake Bay itself to freeze.

Albermale-Pamlico Sound in North Carolina is the second largest estuary in the United States. The average temperature is 43°F (6°C) in winter and 86°F (30°C) in summer. The estuary is cut off from the Atlantic Ocean by the Outer Banks, a long chain of narrow islands. Most of the water in Puget Sound is **brackish** rather than salt water. The estuary has an average depth of 13 feet (4 meters).

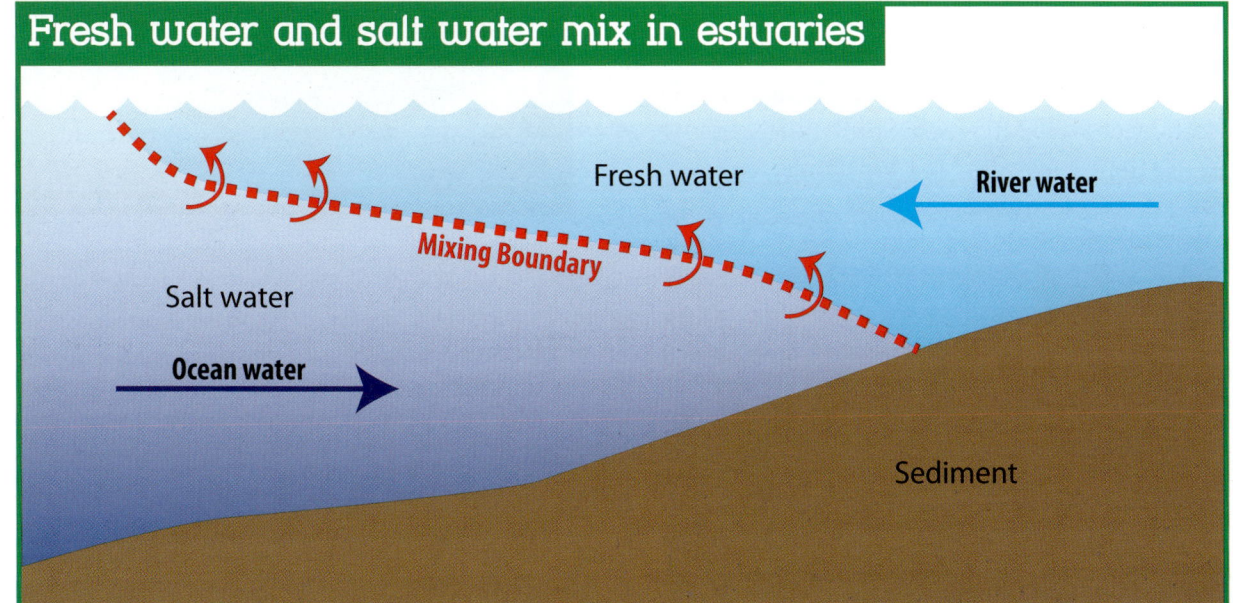

Fresh water and salt water mix in estuaries

11

# Types of Estuaries

There are many kinds of estuaries. The type of estuary that forms depends on the geography of the area.

## Drowned River Valleys

Most drowned river estuaries were formed between 6,000 and 15,000 years ago, at the end of the last **ice age**. Melting ice caused sea levels to rise up to 430 feet (131 m). As coastal valleys were flooded, new estuaries were created.

## Lagoon Estuaries

Barriers formed by islands or narrow strips of land separate these estuaries from the ocean. These estuaries are often less than 16 feet (5 m). They are rarely more than 33 feet (10 m) deep.

## Fjord Estuaries

Fjord estuaries occur in valleys eroded by glaciers. Fjord waters can be more than 1,600 feet (488 m) deep. Above the water, the steep cliff faces of fjords can be up to 4,600 feet (1,402 m) high. Some fjords are more than 100 miles (161 km) long.

## Tectonic Estuaries

**Tectonic** estuaries are the result of land movement caused by earthquakes, volcanoes and other tectonic activity. San Francisco Bay is a tectonic estuary. Earthquakes occur frequently in the San Francisco area.

Ecosystems | Estuaries

# Estuary Features

## Barrier Beaches

These long, narrow strips of sand run **parallel** to the coastline. Barrier beaches are separated from the mainland by a lagoon. Barrier beaches protect the mainland from storms and **erosion** by buffering ocean waves. Behind these beaches, the lagoons are calm and not as affected by wave action.

## Salt Marshes

Salt marshes are common in estuaries. The soil is made of **peat** and deep mud. Salt marshes provide habitats for many plants, fish, birds, and other animals. Like barrier beaches, salt marshes protect shorelines from erosion and flooding.

## Mangrove Swamps

Mangrove trees grow in estuaries along tropical coastlines near the equator. Their large root systems help protect the land from soil erosion and the effects of storms. They also provide feeding and nesting areas for many species.

# Life in Estuaries

Estuaries are some of the most productive ecosystems on Earth. They are home to many different species of fish, birds, **crustaceans**, marine reptiles, and other animals. These organisms depend on each other for the energy they need to survive. This energy transfers between organisms when one organism eats another. Scientists call this transfer of energy between organisms a food chain.

## Producers

The plants and plant-like organisms found in estuaries are the first link in the food chain. Producers absorb energy from the Sun and convert it into usable forms of energy, such as sugar. This process is called photosynthesis. Producers found in estuaries include seagrasses, a wide variety of land and water plants, algae, **plankton**, and bacteria.

## Primary Consumers

The animals that rely on producers as a food source are called primary consumers. Examples of primary consumers found in estuary ecosystems include small fish, birds, and filter feeders, such as crabs and mussels.

## Secondary and Tertiary Consumers

### Estuaries Energy Pyramid

The transfer of energy in an ecosystem begins with producers and moves up the energy pyramid to the tertiary consumers. Organisms at each level of the pyramid receive energy from the organisms in the level below them.

Outside of the pyramid are the organisms called decomposers. They break down the dead and decaying **organic** matter left behind when plants and animals die. For this reason, decomposers receive energy from organisms in all levels of the energy pyramid.

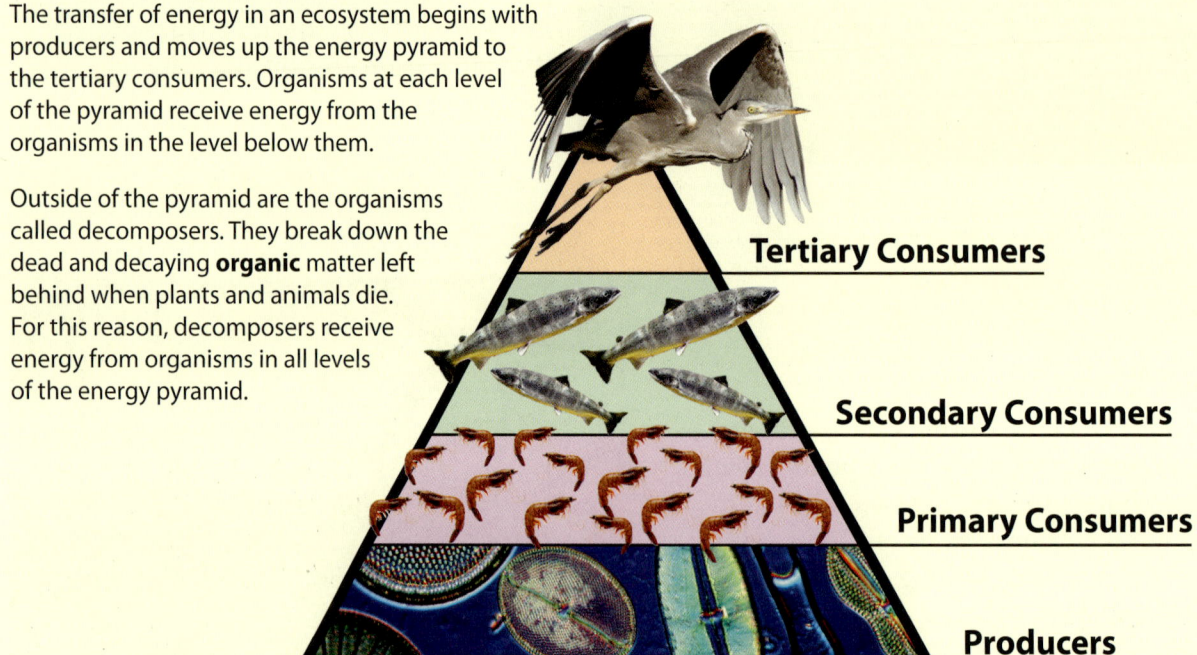

14

Ecosystems | **Estuaries**

## Estuaries Food Web

Another way to study the flow of energy through an ecosystem is by examining food chains and food webs. A food chain only shows how a producer feeds a primary consumer, which then feeds a secondary consumer, and so on. However, most organisms feed on many different food sources. This practice causes food chains to interconnect, creating a food web.

In this example, the **red** line represents one food chain from the algae to the mussel and the heron. The **blue** line from the plankton to the shrimp, crab, and then heron forms another food chain. These food chains connect at the heron, but they also connect in other places. The heron feeds on shrimp, and the shrimp may also eat algae. This series of connections forms a complex food web.

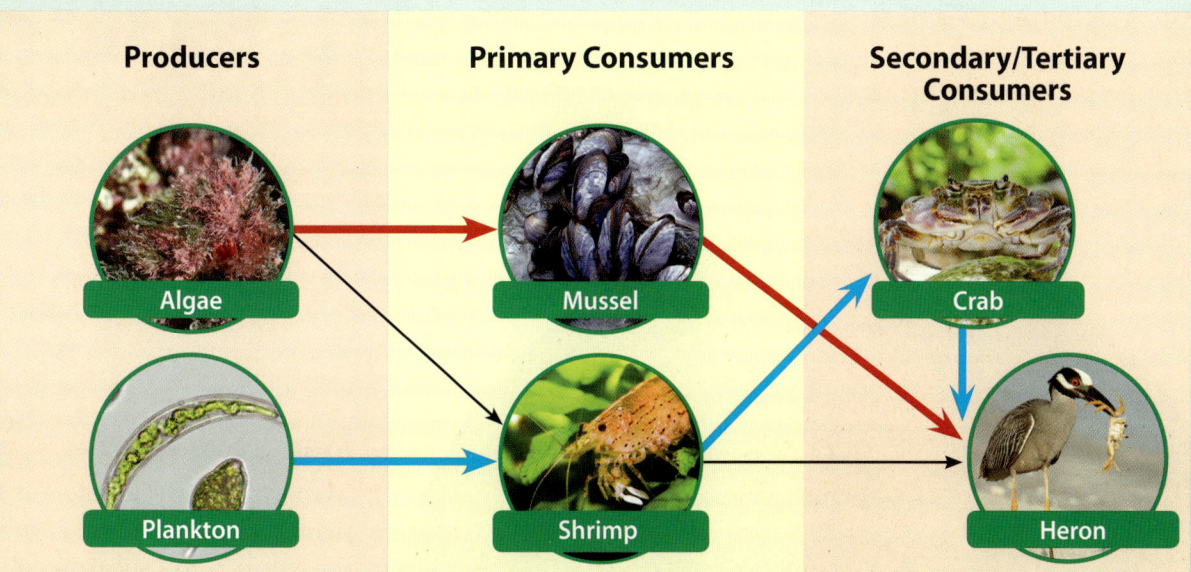

Secondary consumers feed on both producers and primary consumers. In estuaries, secondary consumers include small fish, small birds, and reptiles. Crabs, large fish, and some mammals are also secondary consumers. Larger birds and mammals are called tertiary consumers. Tertiary consumers feed on secondary and primary consumers.

## Decomposers

Crabs, mussels, clams, and many types of bacteria are the main decomposers of estuary ecosystems. These organisms consume dead and decaying organic materials, such as dead fish and plants. Decomposers speed up the process of breaking down these materials. Decomposers are eaten by both secondary and tertiary consumers.

# Estuary Plants

Salt is harmful to most plants. Estuary plants must be able to survive in the salty conditions caused by the mixture of salt water and fresh water.

## Seagrass

There are about 60 different types of seagrasses. They require a great deal of light. Seagrasses are found in estuaries in tropical and temperate regions. They grow in both small patches and huge meadows.

The weedy sea dragon often lives in seagrasses found in estuaries on the southern coast of Australia.

## Mangrove Trees

Mangroves are salt-tolerant plants. They are found along the tropical and subtropical coasts of Africa, Asia, Australia, and South and North America. Mangroves provide food and shelter to crustaceans, fish, birds, and many other types of marine life. Mangroves are also important in some coral reef ecosystems.

The tangled roots of mangrove plants are exposed above the water and are able to filter salt.

Ecosystems | **Estuaries**

### Eco Facts

Some species of mangrove have adapted to block salt from getting into their roots. If any salt does get in, the mangrove directs the salt to specific leaves, which the tree will then drop. This protects the rest of the tree.

## Zostera

Also known as marine eelgrass, this plant can be found beneath the water or partly floating in estuaries. It is an important part of the coastal ecosystem, forming a habitat for many species. Young fish find shelter in the grass as they grow, and mussels attach to the leaves. When the grass washes up on the beach, it provides a habitat for many types of insects.

## Cordgrass

Cordgrass is also called spartina. There are 14 species, with most of these found on the Atlantic coasts of North and South America. It is also found in western and southern Europe. Cordgrass forms thick colonies and grows quickly. It can grow as tall as 6.5 feet (2 m).

Cordgrasses are able to release the salt that they have absorbed from salt water.

# Estuary Mammals

Mammals are warm-blooded animals that are covered with hair or fur and breathe through lungs. Marine mammals, such as dolphins and whales, no longer have much hair, though they still have some. Most marine mammals have a thick layer of fat, called blubber, between their skin and muscles. Blubber keeps these animals warm in cold waters.

## Manatees

Manatees are also known as sea cows. The three species of manatee are the West African, the Amazonian, and the West Indian manatee. Manatees are primarily **herbivores**, but they will sometimes eat fish or **invertebrates**. Manatees live in waters no cooler than 70°F (21°C). Though they can hold their breath up to 20 minutes, they generally surface every few minutes to breathe.

> Manatees can grow up to 13 feet (4 m) long and can weigh up to 1,300 pounds (590 kg).

## River dolphins

River dolphins live in Asia and South America. Four species live in freshwater rivers. The La Plata dolphin is the only species that lives in salt water estuaries and marine environments close to shore. La Plata dolphins are bottom feeders and eat many different species of fish, as well as shrimp, squid, and octopus. They are a small type of dolphin. They grow to about 6 feet (1.8 m) long and may weigh up to 90 pounds (41 kg).

> Each year, more than 1,000 La Plata dolphins are killed. They become tangled in fishing nets and then drown.

Ecosystems | **Estuaries**

### Harbor Seals

Harbor seals are mammals that live along coastlines and in estuaries in the Arctic and temperate regions of the northern **hemisphere**. When in cold water, a harbor seal's skin may be the same temperature as the water but its blubber helps it maintain a warm core body temperature. Harbor seals rest and feed in estuaries and harbors. They feed on many different types of fish, as well as shrimp, squid, and crabs.

| Harbor seals can dive as deep as 984 feet (300 m).

### Sea Otters

Sea otters live along the coasts and estuaries of the northern Pacific ocean. They can walk on land but spend most of their time in water. Adults weigh between 30 and 100 pounds (14 to 45 kg). Instead of blubber, sea otters have a very thick fur coat. Sea otters dive underwater to hunt. They mostly feed on clams, snails, sea urchins, and crustaceans. They may also eat certain kinds of fish. Although sea otters can hold their breath for about five minutes, dives of around one minute are more typical.

| Sea otters must keep their fur very clean in order to stay warm in the water.

### Eco Facts

The Amazonian manatee only lives in the Amazon River and the smaller rivers that feed into it.

19

# Estuary Birds, Reptiles, Fish, and Crustaceans

Estuary birds often migrate, or travel, to warmer places in winter. Salmon only live in estuaries while migrating to other locations. Other organisms, such as some crocodiles, may spend their entire lives in estuaries.

## Great Blue Heron

The Great Blue Heron is the largest heron species in North America. Great Blue Herons are tall and able to stand and feed in deeper waters than other wading birds. The Great Blue Heron spears prey with its long bill, then swallows its catch whole. Herons mostly eat fish, but they also feed on crabs, shrimp, insects, and small birds.

Herons feed in the water of estuaries, but they often make their nests on islands or in swamps with large trees.

There are around 200,000 to 300,000 saltwater crocodiles in the world.

## Estuarine Crocodile

More commonly known as the saltwater crocodile, this species is the largest of all living reptiles. Saltwater crocodiles may be found in river and estuary systems throughout India, southeast Asia, Indonesia, and northern Australia. They may even be found in the open ocean. Saltwater crocodiles can grow up to 17 feet (5.2 m) long and weigh up to 1,000 pounds (454 kg). Saltwater crocodiles are apex **predators**. This means they are at the top of their food chain. They hunt by sneaking up on their prey. Once they are close enough, they will make a sudden attack, catching their prey by surprise. They will then drag their prey underwater to drown it.

Ecosystems | **Estuaries**

### Eco Facts

In Alaska and the Pacific Northwest, salmon support other animals, such as bears, otters, and birds. These animals feed on the salmon during the salmon's mating season.

## Salmon

Most salmon are born in fresh water but live their lives in the ocean. As adults, ocean-living salmon return to fresh water to spawn, or lay their eggs. Once they spawn, the salmon die. When the eggs hatch, the young salmon will live in the freshwater river until they are ready to travel to the ocean. Estuaries provide food and protection. Estuaries also allow young salmon to develop before they migrate to the ocean.

Salmon migrations often involve swimming up river estuaries and leaping out of the water to get past waterfalls and other obstacles.

## Crustaceans

In estuary systems, there are many kinds of crustaceans. These include different species of shrimp, crab, and crayfish. Many of the crustacean species that inhabit estuary systems feed on the detritus, or waste, of other organisms. This may include decaying plant materials and dead fish. Some crustaceans actively hunt prey. For example, the blue crab, which inhabits the Delaware Bay and the eastern coast of the United States, preys upon clams, worms, and shrimp.

The blue crab fishery is the most valuable fishery in the Delaware Bay.

21

# Estuaries in Danger

Many of the world's largest cities are located on estuaries. Buenos Aires, capital of Argentina, and Montevideo, capital of Uruguay, are located along the shores of the Río de la Plata estuary, directly across the river from each other. New York City is located at the mouth of the Hudson River. San Francisco, Seattle, Boston, and New Orleans are all located on estuaries. More than half the population of the United States lives within 100 miles (161 km) of the coast.

Estuaries are threatened by many human activities, including sewage dumping, coastal settlement, and land clearance. Estuaries are also affected by events far upstream. Agricultural chemicals, industrial waste, and other pollutants can enter rivers far inland, eventually making their way to estuaries. Pollutants such as **pesticides**, plastics, and **heavy metals** have been found in estuary ecosystems. They can cause severe damage to many estuary organisms. Frogs, a type of estuary **amphibian**, are very sensitive to pollution. Frog populations around the world have been decreasing greatly over the years due to pollution and other human activities. Overfishing is another major problem for estuary systems. Chesapeake Bay's oyster population has almost been wiped out by overfishing.

## Timeline of Human Activity in Estuaries

**4th century BC** — A chain is set across the entrance to the Golden Horn in Constantinople, modern day Istanbul, to protect the estuary from enemy ships.

**AD 711** — Ostia Antica is founded on the estuary of the Tiber River. It it the main seaport for the Roman Empire. Over the years, **silt** filled the river. Ostia Antica is now 1.8 miles (3 km) from the sea.

**1000** — Viking sailors from Greenland explore the St. Lawrence estuary area.

**1497** — Vasco da Gama sails from the Tagus estuary in Spain around the southern tip of Africa, discovering the sea route to India.

**1534** — French explorer Jacques Cartier becomes the first European to map the St. Lawrence estuary. He names the area Canada.

22

Ecosystems | **Estuaries**

Boston Harbor is an estuary and part of Massachusetts Bay. It was one of the most polluted bodies of water in the United States until a cleanup project began in 1985.

The Forth Bridge is built across Scotland's Firth of Forth estuary. It is designed to carry trains.

Construction begins in the Netherlands to protect the Rhine-Meuse-Scheldt estuaries from flooding.

Scientists declare the Chinese river dolphin, or baiji, extinct.

**1579** | **1890** | **1939** | **1958** | **1982** | **2006**

England's Sir Francis Drake lands near San Francisco Bay during his sailing voyage around the world.

The Battle of the River Plate, the first naval battle of World War II, takes place in the Río de la Plata estuary.

Construction is completed on the Thames Barrier. It protects London from flooding by storm surges from the Thames estuary.

23

# Science in Estuaries

Many sea turtles lay their eggs on estuary shores. During this time, scientists often work at these sites to protect the turtles from predators.

Estuaries and their surrounding coastlines often suffer from flooding by storm surges from the sea. In 1958, the Netherlands government began construction on a series of dams, dikes, levees, and storm surge barriers called the Delta Works to protect the Rhine-Meuse-Scheldt estuaries. The project was completed in 2010. Higher barriers soon may be needed if sea levels rise due to climate change. In 1982, the Thames Barrier near London was completed. This flood barrier protects London from storm surges in the Thames estuary. If there is a flood threat, the barrier only needs to be raised at high tide. It can be lowered at low tide to release the water that has backed up behind the barrier.

Ecosystems | **Estuaries**

## Marine Scientists

Estuaries are important indicators of climate change, pollution levels, water quality, and the sustainability of wildlife populations. In estuaries, scientists identify how many species there are in an ecosystem and how they interact. By doing so, marine scientists may learn more about the effects of human activities, such as fishing and land development. Marine scientists also use satellite images to track and measure pollution levels. As estuaries are ecosystems that include freshwater environments, saltwater environments, and land, they are especially important for understanding the effect humans are having on the biosphere.

In 2010, satellite images of the Deepwater Horizon oil spill helped scientists track the oil as ocean currents pushed it toward the coast of the Gulf of Mexico.

## Wildlife Research

Scientists use radio tagging to learn about animals. Researchers catch animals, such as sea otters, and attach electronic tags to them. They may also weigh the animal, take a blood sample, and perform other tests. They then release the animal back into nature. The tags then broadcast radio signals to research stations, which help track the animals' movements. Using the tracking tags, the scientists may try to catch the same animal again to check its health. Tracking animals helps scientists learn more about an animal's habitat. It also helps them find ways to protect endangered animals and threatened environments.

### Eco Facts

Scientists research the migrations of herons to better understand wildlife living in estuaries. Birds are captured, fitted with a distinctive band, then released. Many herons fly south in winter. Banded birds from Canada have been identified as far south as Mexico, Honduras, and Cuba.

25

# Working in Estuaries

From working with wildlife to developing technology and research equipment, there are many exciting and challenging careers in estuaries.

Many jobs involving the study of estuaries require a background in biology and chemistry. Before considering a career in marine sciences, it is important to research the options, learn about the educational requirements, and get hands-on experience with professionals at coastal, estuary, or marine centers.

Ecosystems | Estuaries

## Marine Ecologist

### Duties
Studies wildlife at the individual, as well as community, level in marine ecosystems

### Education
Bachelor's, master's, or doctoral degree in marine science

### Interests
The environment, biology, ecology, oceanography

Marine ecologists analyze organisms and how they are affected by pollution and other factors. They often study entire populations of a species and how the species interacts with its environment.

## Other Estuary Jobs

### Biologist
Studies the animals that live in estuaries and how they interact with each other and their environments

### Oceanographer
Studies the physical properties of the ocean, including ocean waves, currents, and geology

### Fisheries Officer
Ensures the protection of sensitive coral reef ecosystems by monitoring fishing practices and their effects on the environment

### Marine Photographer
Uses photographic equipment to document the ocean and the many organisms for which it provides a habitat

## Joel Hedgpeth

Joel Hedgpeth (1911–2006) was a marine biologist, author, and environmentalist. Hedgpeth was an expert on bottom-dwelling marine organisms known as sea spiders. His research also led to the California freshwater shrimp being declared an endangered species.

Hedgpeth's love of marine life led him to become a marine biologist. He received his doctoral degree in marine biology from the University of California, Berkeley, in 1952. As a professor, Hedgpeth taught at the University of Texas, the Scripps Institution of Oceanography, the University of the Pacific, as well as Oregon State University.

During his career, Hedgpeth worked with many other marine researchers, including Edward F. Ricketts, a well-known researcher of marine life. Their work together focused on the coasts of California, Oregon, and Washington.

Hedgpeth loved the natural world. In the 1960s, he campaigned to stop the building of a nuclear power plant in California. He also was influential in the west coast environmental movement in the 1970s. Hedgpeth retired in 1973 but continued to write and publish until his death.

# Underwater Viewer

Some estuary organisms, such as crocodiles, have special membranes that allow them to open their eyes underwater. In this activity, you will build an underwater viewer. This viewer will help you to see beneath the surface of the water.

## Materials

Scissors

Plastic Bottle

Clear Plastic Wrap

Rubber Band

1. With an adult's help, use the scissors to cut the bottom and top from the plastic bottle. Try not to leave any sharp edges on the bottle. Recycle the scraps. A clear plastic bottle will allow more light in to make viewing better. Also, the wider the bottle, the wider the view.

2. Cut a piece of plastic wrap that is large enough to cover the widest end of the bottle and to go up its side about 2 inches (5 cm).

3. Use this plastic wrap to cover the widest hole. Secure the wrap to the bottle using the rubber band.

4. Take the underwater viewer to a nearby shallow body of water, such as a stream or creek. Being careful not to fall in the water, hold the viewer in the water so that only the part wrapped in plastic is underwater. Look through the top of the viewer. Do you see any plants or animals? Record your findings.

# Create a Food Web

Ecosystems | Estuaries

Use this book, and research on the Internet, to create a food web of estuary ecosystem producers and consumers. Start by finding at least three organisms of each type—producers, primary consumers, secondary consumers, and tertiary consumers. Then, begin linking these organisms together into food chains. Draw the arrows of each food chain in a different color. Use a **red** pen or crayon for one food chain and **green** and **blue** for the others. You should find that many of these food chains connect, creating a food web. Add the rest of the arrows to complete the food web using a pencil or **black** pen.

Once your food web is complete, use it to answer the following questions.

1. How would removing one organism from your food web affect the other organisms in the web?

2. What would happen to the rest of the food web if the producers were taken away?

3. How would decomposers fit into the food web?

## Sample Food Web

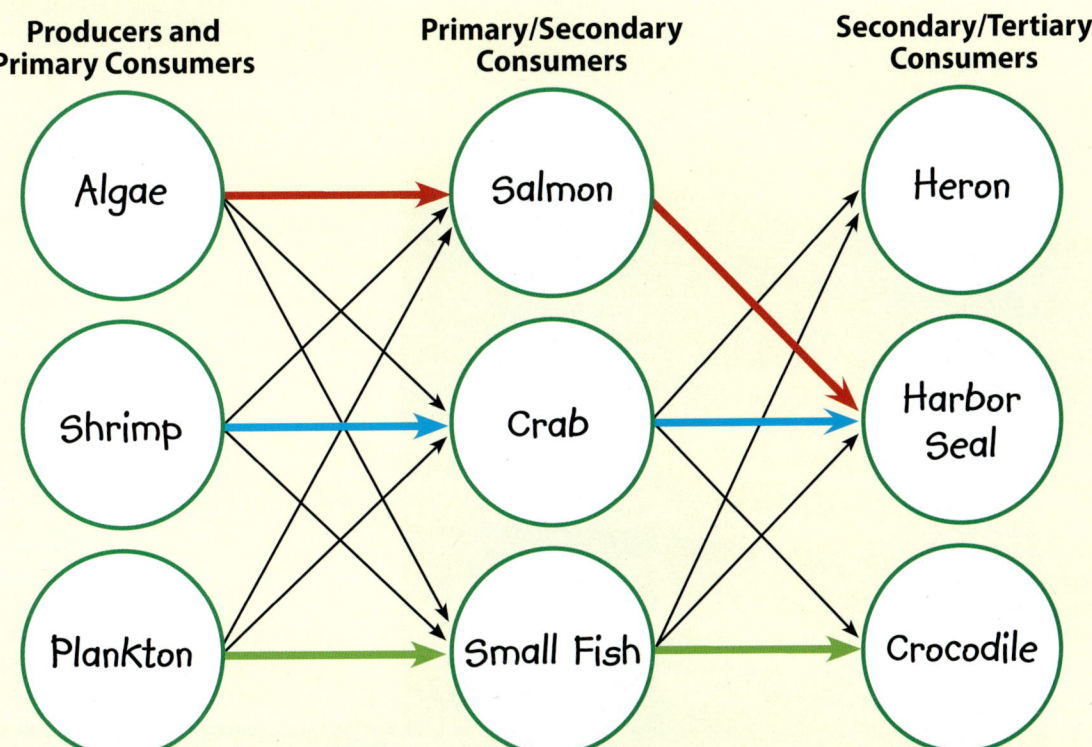

# Eco Challenge

1. What are barrier beaches?
2. What is the area of the Amazon estuary?
3. What is the world's largest estuary?
4. What is the average January temperature in the Ob estuary?
5. What is a fjord estuary?
6. What is the largest estuary in the United States?
7. How many species of manatee are there?
8. What is zostera also known as?
9. Name some of the ways in which estuary ecosystems are endangered by human activity.
10. What do the Delta Works prevent from flooding?

## Answers

1. Narrow strips of sand that run parallel to the coastline and are separated from it by a lagoon
2. Approximately 7,800 square miles (20,202 sq. km)
3. The Gulf of St. Lawrence
4. −18°F (−28°C)
5. An estuary in a valley eroded by glaciers
6. Chesapeake Bay
7. Three
8. Marine eelgrass
9. Sewage, coastal settlement, land clearance, pesticides, heavy metals, overfishing
10. The Rhine-Meuse-Scheldt estuaries

# Ecosystems | Estuaries

# Key Words

**amphibian:** a cold-blooded vertebrate, such as a frog, that lives on land and in water

**brackish:** water with more salt than fresh water, but not as much as seawater

**crustaceans:** organisms with a hard shell to protect their boneless, soft bodies

**ecosystem:** a community of living things sharing an environment

**equator:** an imaginary line drawn around Earth's center

**erosion:** the wearing away of material, such as earth or rock

**heavy metals:** metallic elements that are toxic to organisms in low amounts

**hemisphere:** one half of Earth

**herbivores:** animals that only eat plants

**ice age:** periods in Earth's history when ice covered much of the planet

**invertebrates:** animals without backbones, or spines.

**lagoon:** a shallow body of water separated from a larger body by reefs or small islands

**organic:** made of living things

**organisms:** living things

**parallel:** two things side by side with the same distance always between them

**peat:** a soil-like material composed of vegetable matter

**pesticides:** chemicals used to kill insects

**plankton:** tiny organisms that float in fresh water or salt water

**predators:** animals that hunt other animals for food

**silt:** fine sand that builds up on the bottoms of rivers, lakes, and other bodies of water

**species:** a group of similar animals that can mate together

**tectonic:** land movement caused by earthquakes, volcanoes, and similar activity

**tropical:** relating to the warm areas near the equator

# Index

Amazon  7, 8, 11, 18, 19, 30

Chesapeake Bay  7, 8, 11, 22, 30

Delta Works  24, 30

harbor seal  19, 29

lagoon  4, 12, 13, 30

mammals  15, 18, 19
manatee  18, 19, 30
mangroves  13, 16
marine biologist  27

Ob River  6, 9, 10, 30
oceanographer  27

Río de la Plata  7, 8, 22, 23
river dolphin  18, 23

St. Lawrence, Gulf of  6, 10, 22, 30
sea otters  19, 25
seagrasses  14, 16

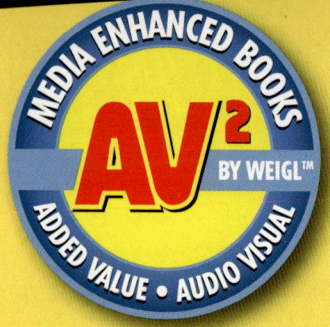

# Log on to www.av2books.com

AV² by Weigl brings you media enhanced books that support active learning. Go to www.av2books.com, and enter the special code found on page 2 of this book. You will gain access to enriched and enhanced content that supplements and complements this book. Content includes video, audio, weblinks, quizzes, a slide show, and activities.

## AV² Online Navigation

**Audio**
Listen to sections of the book read aloud.

**Book Pages**
AV² pages directly correspond to pages in the book.

**Video**
Watch informative video clips.

**Key Words**
Study vocabulary, and complete a matching word activity.

**Embedded Weblinks**
Gain additional information for research.

**Quizzes**
Test your knowledge.

**Slide Show**
View images and captions, and prepare a presentation.

**Try This!**
Complete activities and hands-on experiments.

AV² was built to bridge the gap between print and digital. We encourage you to tell us what you like and what you want to see in the future.

## Sign up to be an AV² Ambassador at www.av2books.com/ambassador.

Due to the dynamic nature of the Internet, some of the URLs and activities provided as part of AV² by Weigl may have changed or ceased to exist. AV² by Weigl accepts no responsibility for any such changes. All media enhanced books are regularly monitored to update addresses and sites in a timely manner. Contact AV² by Weigl at 1-866-649-3445 or av2books@weigl.com with any questions, comments, or feedback.